NATIONAL GEOGRAPHIC

Ladders

Welcome to India

AROUND THE WORLD

GENRE Social Studies Article | **Read to find out** about living in villages and cities in India.

WELCOME TO INDIA

by Stephanie Herbek

Namaste! Hello! Welcome to India. India is a large country on the continent of Asia. It has mountains, rivers, and deserts. Over one billion people live here.

Culture is important to people in India. Culture includes food, beliefs, music, art, and language. Culture can be different in small communities and big communities.

> In India, people ride animals alongside cars and trucks. Elephants are a good thing to ride. They have a lot of trunk space!

Small Villages and Big Cities

Most people in India live in small communities called villages. Indian villages have simple homes and wide-open spaces. Many people also live in India's cities. These cities are big, crowded communities. Life in an Indian village can be different from life in an Indian city.

People from **villages** sometimes buy or sell things in cities. People from cities often visit friends in villages. Let's find out more about Indian villages and cities!

Living in a Village

Indian villages are in the country. Most villages are made up of small homes, a few stores, and a post office. The homes are usually built of mud, straw, and stone. Most village homes don't have electricity, running water, or TV. People in Indian villages get water from a **well**.

Many villagers are farmers. They grow crops to eat and sell. Lucky villagers might also own a buffalo. The buffalo gives them milk. It might also live in the house. That's a big pet!

This fishing village is near the water. Check out the colorful boats and homes.

In a village, people wake early. Their days are busy! They pump water at the well, gather wood for the fire, and cook the morning meal. Then most children head off to small village schools. Other children work on their family's farm. When school and work are done, all village children play together.

In this village, children play near homes made of wood and grass.

Living in a City

India also has some of the most crowded cities in the world. Over ten million people live in Mumbai (muhm-BY). This busy city has highways and tall buildings. Most Indians who live in cities live in small apartments. They work in offices and factories. They do not have space to grow their own food, so they buy it in markets.

> In busy Indian cities, people buy their food at outdoor markets.

Most families in Mumbai live in apartment buildings.

Cities in India have parks, museums, restaurants, and hotels. Buses, bicycles, and trains help people get from place to place. People also ride the train from villages to cities and back again.

Check In How is living in a city in India different from living in a village?

GENRE Culture Article

Read to find out about going to school in India.

Going to School in India

by Grant Williams

Children in India get to school in many different ways. In cities, children often ride the bus to school. But some village children have to travel past mountains, rivers, and forests. Imagine riding a boat across a river and tiptoeing along a swaying bridge just to get to school!

> This boy gets a ride to school each day on his father's bike.

Some Indian children learn in unusual places. Schools in India are not always inside a building. Students in small villages can learn outdoors. Students in desert communities sometimes learn inside a tent. Students in busy cities might learn inside a parked bus.

Many children walk to school in India. Others think it's more fun to run!

Coming through! These children get pulled to school in a little cart.

It's the rainy season! These girls have to wade through flooded streets to get to school.

Learning and Playing

Once they get to school, children are ready to learn. They study math, science, and art. They also learn to speak Hindi and English. These are the most important languages in India. Many people in India speak Hindi. Learning English helps Indians talk to people around the world.

> In some Indian classrooms, children sit on the floor instead of at desks.

Children in India exercise during the school day. Many children like to play **cricket** and soccer. Cricket is a sport like baseball. It's played on a field with a ball and a bat. Soccer is a sport played all over the world!

Small chalkboards, or slates, make it easy to practice writing words in English.

At recess, some children play games with marbles.

Wearing a uniform is usually part of going to school in India.

Lunchtime!

What's for lunch in India? Indian farmers grow rice, wheat, peas, and beans. Indian children eat many meals made with these foods. Spices, such as **curry**, are often added to vegetable or meat dishes. Chili peppers also make Indian foods spicy. Most meals are served with rice.

> These children are enjoying their lunch.

Children in India carry their lunches in a metal lunch box called a *tiffin*. Indian people do not eat very much meat. So most children bring rice dishes to school. Some also bring *shahi paneer*. This is a mixture of cheese and vegetables. Indian children also bring water to school. Not all schools have running water.

Kids can carry their lunches to school in tiffins.

Naan bread is round, flat, and very tasty.

This delicious curry is made with spicy vegetables and rice.

No School Today!

In India, children sometimes miss school to celebrate important holidays. Children might go to parties, eat special foods, and spend time with their family and friends.

Diwali (dih-WAH-lee) is one of the most important **festivals** in India. It is celebrated in October or November. Diwali is called "the festival of lights." People light candles. Colorful fireworks light up the night sky. It is a happy time of year.

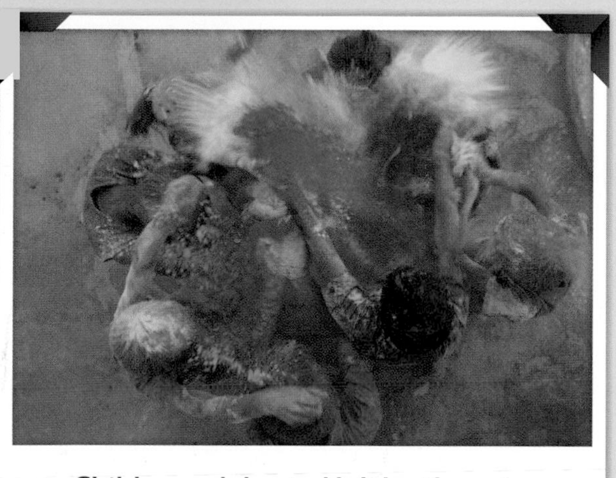

Children celebrate Holi by throwing colorful powders at their friends.

Another important festival in India is called Holi (HOH-lee). Holi celebrates the beginning of spring. It lasts for five days. Holi means "the festival of colors." People pray and eat special foods during Holi. They also throw colorful powders at each other. What fun!

Check In What are some of the different ways students get to school in India?

GENRE Folk Tale

Read to find out how rumors get started.

The Foolish, Timid Rabbit

A TRADITIONAL INDIAN FOLK TALE

retold by Jenny Loomis
illustrated by David Mottram

Many children's stories have lessons about life. Jataka tales are one way that children in India learn **morals**. Morals are lessons about right and wrong. Jataka tales are very old. They come from Buddhism. This is a religion that started in India 2,500 years ago. The moral in this tale is about the danger of telling **rumors**. As you read this folk tale, think about the lessons that it teaches. Do they sound familiar to you?

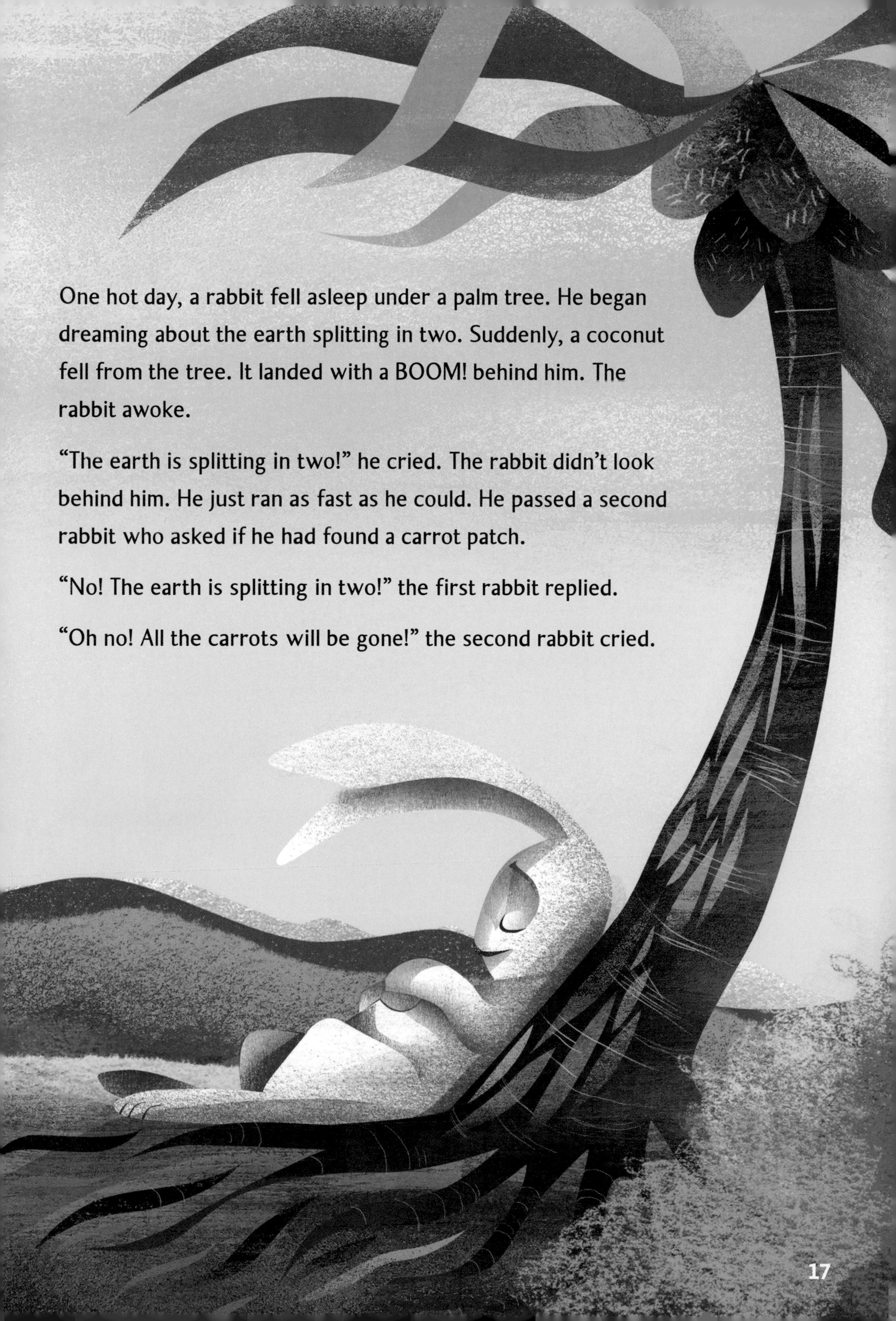

One hot day, a rabbit fell asleep under a palm tree. He began dreaming about the earth splitting in two. Suddenly, a coconut fell from the tree. It landed with a BOOM! behind him. The rabbit awoke.

"The earth is splitting in two!" he cried. The rabbit didn't look behind him. He just ran as fast as he could. He passed a second rabbit who asked if he had found a carrot patch.

"No! The earth is splitting in two!" the first rabbit replied.

"Oh no! All the carrots will be gone!" the second rabbit cried.

The second rabbit ran to catch up with the first rabbit. The two of them raced through a field. They passed rabbit after rabbit and told each one that the earth was splitting in two. Soon, hundreds of rabbits were running.

They ran past a monkey. He was eating a banana in a tree. “Where are you going?” the monkey asked. The rabbits told the monkey that the earth was splitting in two.

“What will happen to my bananas?” cried the monkey. He swung down from the tree to join them.

The rabbits and the monkey raced past an elephant. He asked about the fuss. The monkey explained that the earth was splitting in two. The elephant said, "How awful! This can't be good for the plants and trees!" He joined the other animals. Then the animals passed a deer snacking on some moss.

"What on earth is going on?" asked the deer.

"The earth is splitting in two!" answered the elephant.

"Oh my! I hope the moss won't get hurt!" The deer ran to the front of the line of animals.

The lion saw the long line of running animals from the top of a hill. He was worried. He ran to the foot of the hill and let out a roar. The animals stopped. They knew not to disobey him.

"What is going on?" asked the lion.

"The earth is splitting in two!" the animals yelled together.

"How do you know this?" asked the lion.

“I think the elephant knows,” said the deer.

“Check with the monkey,” said the elephant.

“The rabbits told me!” said the monkey. The rabbits looked at one another.

“Who started all of this?” roared the lion.

The first rabbit quietly raised his paw. He said that he learned first that the earth was splitting in two. He described the sound he heard while napping under the tree.

“Show me where this happened. The rest of you wait here,” ordered the lion.

The rabbit and lion ran back to the palm tree.

"This is where it happened, King Lion!"

The lion looked at the area near the palm tree. Then he saw the coconut.

"Foolish rabbit, the earth isn't splitting in two. The BOOM! you heard was the sound of this coconut falling from its tree. Let's go back to the others and explain what happened."

They returned to the hill. The lion told the animals that the earth was not splitting in two. The animals cheered.

"Next time, be sure a rumor is true before acting on it," ordered the lion.

The animals thought this was a smart plan. They thanked the lion for his help. Without it, they might still be running.

Check In What lesson does this folk tale teach?

Discuss

1. Tell about some of the ways you think the three selections in this book are linked.

2. Do most people in India live in small villages or in big cities? Which place would you rather live? Why?

3. Which Indian festival is similar to a holiday or festival that you celebrate? How is it similar?

4. Do you think the first rabbit in the folk tale acted foolishly? What makes you think that?

5. What do you still wonder about the land and people of India?